LCM

THINGS YOU SHOULD KNOW
(QUESTIONS AND ANSWERS)

By Rumi Michael Leigh

Introduction

I would like to thank you for purchasing this book, *"LCM", things you should know (questions and answers)"*.

This book will help you understand, revise, and have a good general knowledge and understanding of the basics of LCM.

I hope you enjoy it!

Table of Contents

Part 1

1.

What is LCM?

LCM is the Least Common Multiple

Part 2a: LCM of two numbers (Questions)

Find the LCM of the following numbers

a. 2 and 3

b. 3 and 4

c. 4 and 5

d. 5 and 6

e. 6 and 7

f. 7 and 8

g. 8 and 9

h. 9 and 10

i. 10 and 11

j. 11 and 12

Part 2a: LCM of two numbers (Solutions)

Find the LCM of the following numbers

a. 2 and 3

The multiples of 2 are 2, 4, 6, 8, 10, 12, 14, 16, etc.
The multiples of 3 are 3, 6, 9, 12, 15, 18, 21, 24, etc.
The common multiples of 2 and 3 are 6, 12
So, the LCM of 2 and 3 is 6

b. 3 and 4

The multiples of 3 are 3, 6, 9, 12, 15, 18, 21, 24, etc.
The multiples of 4 are 4, 8, 12, 16, 20, 24, 28, 32, etc.
The common multiples of 3 and 4 are 12, 24, ...
So, the LCM of 3 and 4 is 12

c. 4 and 5

The multiples of 4 are 4, 8, 12, 16, 20, 24, 28, 32, 36, 40, etc.
The multiples of 5 are 5, 10,15, 20, 25, 30, 35, 40, 45, etc.
The common multiples of 4 and 5 are 20, 40, ...
So, the LCM of 4 and 5 is 20

d. 5 and 6

The multiples of 5 are 5, 10, 15, 20, 25, 30, 35, 40, 45, 50, 55, 60, ...
The multiples of 6 are 6, 12, 18, 24, 30, 36, 42, 48, 54, 60, ...
The common multiples of 5 and 6 are 30, 60, ...
So, the LCM of 5 and 6 is 30

e. 6 and 7

The multiples of 6 are 6, 12, 18, 24, 30, 36, 42, 48, 54, 60, 66, 72, ...

The multiples of 7 are 7, 14, 21, 28, 35, 42, 49, 56, 63, 70, 77, ...

So, the LCM of 6 and 7 is 42

f. 7 and 8

The multiples of 7 are 7, 14, 21, 28, 35, 42, 49, 56, 63, 70, ...

The multiples of 8 are 8, 16, 24, 32, 40, 48, 56, 64, 72, 80, ...

So, the LCM of 7 and 8 is 56

g. 8 and 9

The multiples of 8 are 8, 16, 24, 32, 40, 48, 56, 64, 72, 80, ...

The multiples of 9 are 9, 18, 27, 36, 45, 54, 63, 72, 81, 90, ...

So, the LCM of 8 and 9 is 72

h. 9 and 10

The multiples of 9 are 9, 18, 27, 36, 45, 54, 63, 72, 81, 90, 99, ...
The multiples of 10 are 10, 20, 30, 40, 50, 60, 70, 80, 90, 100, ...
So, the LCM of 9 and 10 is 90

i. 10 and 11

The multiples of 10 are 10, 20, 30, 40, 50, 60, 70, 80, 90, 100, 110, ...
The multiples of 11 are 11, 22, 33, 44, 55, 66, 77, 88, 99, 110, ...
So, the LCM of 10 and 11 is 110

j. 11 and 12

The multiples of 11 are 11, 22, 33, 44, 55, 66, 77, 88, 99, 110, 121, 132, ...
The multiples of 12 are 12, 24, 36, 48, 60, 72, 96, 108, 120, 132, 144, ...
So, the LCM of 11 and 12 is 132

Part 2b: LCM of two numbers (Questions)

Find the LCM of the following numbers

a. 9 and 12

b. 12 and 14

c. 16 and 18

d. 6 and 9

e. 7 and 11

f. 14 and 18

g. 14 and 16

h. 8 and 12

i. 6 and 8

j. 4 and 6

Part 2b: LCM of two numbers (Solutions)

Find the LCM of the following numbers

a. 9 and 12

The multiples of 9 are 9, 18, 27, 36, 45, 54, 63, 72, 81, 90, 99, 108, ...
The multiples of 12 are 12, 24, 36, 48, 60, 72, 84, 96, 108, 120, ...
The common multiples of 9 and 12 are 36, 72, 108, ...
So, the LCM of 9 and 12 is 36

b. 12 and 14

The multiples of 12 are 12, 24, 36, 48, 60, 72, 84, 96, 108, 120, ...
The multiples of 14 are 14, 28, 42, 56, 70, 84, 98, 112, 126, ...
So, the LCM of 12 and 14 is 84

c. 16 and 18

The multiples of 16 are 16, 32, 48, 64, 80, 96, 112, 128, 144, 160, 176, ...
The multiples of 18 are 18, 36, 54, 72, 90, 108, 126, 144, 162, 180, 198, ...
So, the LCM of 16 and 18 is 144

d. 6 and 9

The multiples of 9 are 9, 18, 27, 36, 45, 54, 63, 72, 81, 90, 99, 108, ...
The multiples of 6 are 6, 12, 18, 24, 30, 36, 42, 48, 54, 60, 66, 72, ...

The common multiples of 6 and 9 are 18, 36, 54, 72, ...

So, the LCM of 6 and 9 is 18

e. 7 and 11

The multiples of 7 are 7, 14, 21, 28, 35, 42, 49, 56, 63, 70, 77, 84, ...
The multiples of 11 are 11, 22, 33, 44, 55, 66, 77, 88, 99, ...
So, the LCM of 7 and 11 is 77

f. 14 and 18

The multiples of 14 are 14, 28, 42, 56, 70, 84, 98, 112, 126, ...
The multiples of 18 are 18, 36, 54, 72, 90, 108, 126, 144, 162, ...
So, the LCM of 14 and 18 is 126

g. 14 and 16

The multiples of 14 are 14, 28, 42, 56, 70, 84, 98, 112, 126, ...
The multiples of 16 are 16, 32, 48, 64, 80, 96, 112, 128, 144, 160, 176, etc.
So, the LCM of 14 and 16 is 112

h. 8 and 12

The multiples of 8 are 8, 16, 24, 32, 40, 48, 56, 64, 72, 80, 88, 96, ...
The multiples of 12 are 12, 24, 36, 48, 60, 72, 84, 96, 108, 120, ...
The common multiples of 8 and 12 are 24, 48, 72, 96, ...
So, the LCM of 8 and 12 is 24

i. 6 and 8

The multiples of 6 are 6, 12, 18, 24, 30, 36, 42, 48, 54, 60, 66, 72, ...

The multiples of 8 are 8, 16, 24, 32, 40, 48, 56, 64, 72, ...

The common multiples of 6 and 8 are 24, 48, 72, ...

So, the LCM of 6 and 8 is 24

j. 4 and 6

The multiples of 4 are 4, 8, 12, 16, 20, 24, 28, 32, 36, 40, 44, ...

The multiples of 6 are 6, 12, 18, 24, 30, 36, 42, 48, 54, 60, ...

The common multiples of 4 and 6 are 12, 24, 36, ...

So, the LCM of 4 and 6 is 12

Part 3a: LCM of three numbers (Questions)

Find the LCM of the following numbers

a. 3, 4, 6
b. 2, 6, 8
c. 3, 6, 8
d. 4, 8, 9
e. 6, 8, 12
f. 4, 8, 12
g. 5, 10, 15
h. 2, 4, 6
i. 5, 10, 12
j. 3, 6, 10

Part 3a: LCM of three numbers (Solutions)

Find the LCM of the following numbers

a. 3, 4, 6

The multiples of 3 are 3, 6, 9, 12, 15, 18, 21, 24, 27, 30, 33, 36, 39, ...
The multiples of 4 are 4, 8, 12, 16, 20, 24, 28, 32, 36, 40, ...
The multiples of 6 are 6, 12, 18, 24, 30, 36, 42, ...
So, the LCM of 3, 4, 6 is 12

b. 2, 6, 8

The multiples of 2 are 2, 4, 6, 8, 10, 12, 14, 16, 18, 20, 22, 24, ...

The multiples of 6 are 6, 12, 18, 24, 30, 36, 42, 48, 54, 60, ...

The multiples of 8 are 8, 16, 24, 32, 40, 48, 56, 64, 72, 80, ...

So, the LCM of 2, 6, 8 is 24

c. 3, 6, 8

The multiples of 3 are 3, 6, 9, 12, 15, 18, 21, 24, 27, 30, 33, 36, 39, ...

The multiples of 6 are 6, 12, 18, 24, 30, 36, 42, 48, 54, 60, ...

The multiples of 8 are 8, 16, 24, 32, 40, 48, 56, 64, 72, 80, ...

So, the LCM of 3, 6, 8, is 24

d. 4, 8, 9

The multiples of 4 are 4, 8, 12, 16, 20, 24, 28, 32, 36, 40, 44, 48, 52, 56, 60, 64, 68, 72, ...

The multiples of 8 are 8, 16, 24, 32, 40, 48, 56, 64, 72, 80, ...

The multiples of 9 are 9, 18, 27, 36, 45, 54, 63, 72, 81, 90, ...

So, the LCM of 4, 8, 9 is 72

e. 6, 8, 12

The multiples of 6 are 6, 12, 18, 24, 30, 36, 42, 48, 54, 60, ...

The multiples of 8 are 8, 16, 24, 32, 40, 48, 56, 64, 72, 80, ...

The multiples of 12 are 12, 24, 36, 48, 60, 72, 84, 96, 108, 120, ...

The common multiples of 6, 8, 12 are 24, 48, ...

So, the LCM of 6, 8, 12 is 24

f. 4, 8, 12

The multiples of 4 are 4, 8, 12, 16, 20, 24, 28, 32, 36, 40, 44, 48, 52, 56......

The multiples of 8 are 8, 16, 24, 32, 40, 48, 56, 64, 72, 80, ...

The multiples of 12 are 12, 24, 36, 48, 60, 72, 84, 96, 108, 120, ...

The common multiples of 4, 8, 12 are 24, 48, ...
So, the LCM of 4, 8, 12 is 24

g. 5, 10, 15

The multiples of 5 are 5, 10, 15, 20, 25, 30, 35, 40,
45, 50, 55, 60, 65, 70, 75, ...
The multiples of 10 are 10, 20, 30, 40, 50, 60, 70,
80, 90, 100, ...
The multiples of 15 are 15, 30, 45, 60, 75, 90, 105,
120, 135, ...
The common multiples of 5, 10, 15 are 30, 60, ...
So, the LCM of 5, 10, 15 is 30

h. 2, 4, 6

The multiples of 2 are 2, 4, 6, 8, 10, 12, 14, 16, 18,
20, 22, 24, ...
The multiples of 4 are 4, 8, 12, 16, 20, 24, 28, 32,
36, 40, ...
The multiples of 6 are 6, 12, 18, 24, 30, 36, 42, 48,
54, 60, ...
The common multiples of 2, 4, 6 are 12, 24, ...

So, the LCM of 2, 4, 6 is 12

i. 5, 10, 12

The multiples of 5 are 5, 10, 15, 20, 25, 30, 35, 40,
45, 50, 55, 60, 65, 70, 75, ...
The multiples of 10 are 10, 20, 30, 40, 50, 60, 70,
80, 90, 100, ...
The multiples of 12 are 12, 24, 36, 48, 60, 72, 84,
96, 108, 120, ...
So, the LCM of 5, 10, 12 is 60

j. 3, 6, 10

The multiples of 3 are 3, 9, 12, 15, 18, 21, 24, 27,
30, 33, 36, ...
The multiples of 6 are 6, 12, 18, 24, 30, 36, 42, 48,
54, 60, ...
The multiples of 10 are 10, 20, 30, 40, 50, 60, 70,
80, 90, ...
So, the LCM of 3, 6, 10 is 30

Part 3b: LCM of three numbers (Questions)

Find the LCM of the following numbers

a. 2, 7, 12

b. 8, 12, 14

c. 9, 12, 15

d. 2, 5, 8

e. 2, 7, 9

Part 3b: LCM of three numbers (Solutions)

Find the LCM of the following numbers

a. 2, 7, 12

The multiples of 2 are 2, 4, 6, 8, 10, 12, 14, 16, 18, ...84, ...
The multiples of 7 are 7, 14, 21, 28, 35, 42, 49, 56, ...84, ...
The multiples of 12 are 12, 24, 36, 48, 60, 72, 84, ...
So, the LCM of 2, 7, 12 is 84

b. 8, 12, 14

The multiples of 8 are 8, 16, 24, 32, 40, ...168, ...

The multiples of 12 are 12, 24, 36, 48, ...168, ...

The multiples of 14 are 14, 28, 42, 56, 70, 84, ...168, ...

So, the LCM of 8, 12, 14 is 168

c. 9, 12, 15

The multiples of 9 are 9, 18, 27, 36, ...180, ...

The multiples of 12 are 12, 24, 36, ...180, ...

The multiples of 15 are 15, 30, 45, 60, ...180, ...

So, the LCM of 9, 12, 15 is 180

d. 2, 5, 8

The multiples of 2 are 2, 4, 6, 8, 10, 12, ...40, ...

The multiples of 5 are 5, 10, 15, 20, 25, ...40, ...

The multiples of 8 are 8, 16, 24, 32, 40, ...
So, the LCM of 2, 5, 8 is 40

e. 2, 7, 9

The multiples of 2 are 2, 4, 6, 8, 10, ...126, ...
The multiples of 7 are 7, 14, 21, 28, ...126, ...
The multiples of 9 are 9, 18, 27, ...126, ...
So, the LCM of 2, 7, 9 is 126

Conclusion

Thank you once again for purchasing this book. I hope it has helped you in your journey to understand LCM.

Please, if you learnt something from this book, I would like you to leave a review. It'd be appreciated.

Thank you.